# Why Is Earth Round?

Patricia J. Murphy

The Rosen Publishing Group's
**PowerKids Press**™
New York

*For Peace on Earth*

Published in 2004 by The Rosen Publishing Group, Inc.
29 East 21st Street, New York, NY 10010

Copyright © 2004 by The Rosen Publishing Group, Inc.

All rights reserved. No part of this book may be reproduced in any form without permission in writing from the publisher, except by a reviewer.

First Edition

Editors: Frances E. Ruffin, Natashya Wilson
Book Design: Danielle Primiceri
Layout: Nick Sciacca

Photo Credits: Cover, pp. 4, 7, 8, 15, 19 (lower left) © PhotoDisc; pp. 11, 16, 19 © Nova Development Corporation; p. 12 © Photri Microstock, Inc.; p. 20 © Roger Ressmeyer/CORBIS.

Murphy, Patricia J., 1963–
Why is earth round? / by Patricia J. Murphy.— 1st ed.
   p. cm. — (The library of why?)
Includes bibliographical references and index.
ISBN 0-8239-6236-9 (lib. bdg.)
1. Earth—Juvenile literature. [1. Earth.] I. Title. II. Series.
QB631.4 .M86 2003
525—dc21
                                                    2001006654

Manufactured in the United States of America

# Contents

| | | |
|---|---|---|
| 1 | What Is Earth? | 5 |
| 2 | What Is Earth Made Of? | 6 |
| 3 | How Was Earth Formed? | 9 |
| 4 | Why Is Earth Round? | 10 |
| 5 | Why Did People Believe Earth Was Flat? | 13 |
| 6 | How Does Earth Get Around? | 14 |
| 7 | How Is Earth More than a Sphere? | 17 |
| 8 | What Makes Earth's Seasons? | 18 |
| 9 | Why Do Scientists Study Earth? | 21 |
| 10 | How Can You Find Your Way on Earth? | 22 |
| | Glossary | 23 |
| | Index | 24 |
| | Web Sites | 24 |

# What Is Earth?

Planet Earth is our home. We breathe its air and drink its water. We grow seeds in its soil and build houses on its surface. Earth is the fifth-largest planet in our solar system. It is the third-closest planet to the Sun. The Sun gives Earth heat and light. Earth is surrounded by layers of gases called the **atmosphere**. Earth's atmosphere makes Earth unlike any other planet. Its oxygen lets us breathe. Its **ozone** layer protects us from the Sun's harmful rays. Without Earth's atmosphere, nothing could live on Earth!

◀ *This is how Earth appears from outer space. The blue areas are water. Earth is 70 percent water and 30 percent land.*

# What Is Earth Made Of?

Earth is made of three main layers. The top layer is called the crust. It is made of rocks, such as marble, granite, and shale. The crust is always changing. It is broken into parts called plates. The plates float on Earth's middle layer, the mantle. Part of the mantle is hot, solid rock made of metals, such as iron and magnesium. Some of Earth's mantle is **magma**, or liquid rock. Far below the mantle and the magma is Earth's core. The innermost core is the center of Earth. It is made of the metals iron and nickel. The outside of the core is melted iron and nickel.

*This picture shows the three layers of Earth. The crust is thicker under Earth's land and thinner under its oceans.* ▶

**Crust** →

**Core** ↓

**Mantle** ↑

# How Was Earth Formed?

Most scientists believe Earth began as part of a huge cloud of dust and gas. The cloud spun around and around. **Gravity** pulled the dust and gases closely together. Inside the cloud, **temperature** and pressure grew higher and higher. The center of the cloud became the Sun! The rest of the cloud separated into parts. As the parts moved around, they hit one another. Some of the parts stuck together and grew larger. They became Earth, the other planets, moons, and **meteoroids**.

◀ *Billions of years ago, Earth may have formed inside a huge cloud of dust and gas, like this one.*

# Why Is Earth Round?

Gravity causes Earth's round shape. Gravity is a force that pulls everything on Earth toward Earth's center. This makes Earth round, but Earth is not perfectly round. It is wider through the middle than it is through its North and South Poles. Scientists believe that Earth's spinning on its **axis** causes this less-than-perfect roundness. The axis is a pretend rod that runs through Earth's poles. As Earth spins on its axis, its middle pulls away from its center. This makes Earth wider around its middle.

*Earth measures 24,902 miles (40,076 km) around its middle, but only 24,860 miles (40,008 km) around its poles.*

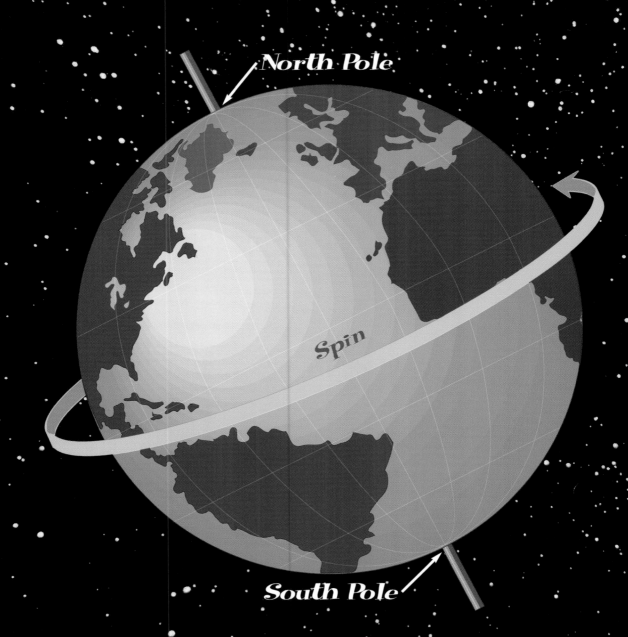

# Why Did People Believe Earth Was Flat?

Early people who didn't know a lot about science thought Earth was flat. They thought they could fall off Earth's edge. Maps were drawn on flat paper, which made people think Earth was flat. In the sixth century B.C., a man named Pythagoras noticed that ships became smaller and disappeared as they sailed away. He thought Earth was round, like the Sun and the Moon. In 1492, explorer Christopher Columbus sailed from Spain to the Bahamas. Most people believed Earth was round when he did not fall off an edge!

◀ *Scientist Galileo Galilei (1564–1642) was one of the first to say that Earth circled the Sun, rather than that the Sun circled Earth.*

# How Does Earth Get Around?

Earth never stops moving. When Earth spins on its axis, day turns to night. It takes Earth about 24 hours at 1,500 feet per second (457 m/s) to complete one spin. Earth circles the Sun at 66,600 miles per hour (107,182 km/h). It takes Earth one year, or 365 days, to complete this orbit of the Sun. As Earth moves around the Sun, Earth's seasons change. Earth also travels around the Milky Way **galaxy**. As the Sun orbits the galaxy at 155 miles per second (250 km/s), it pulls along its circling planets, including Earth.

*This picture shows the Milky Way galaxy. A trip around the Milky Way takes 250 million years!* ▶

# How Is Earth More than a Sphere?

When we talk about parts of Earth, we often talk about the Northern, Southern, Eastern, and Western **Hemispheres**. "Hemisphere" means half a sphere, or ball. The Northern and Southern Hemispheres are separated by a make-believe line around Earth's middle. This line, the **equator**, is halfway between the North and South Poles. Earth is split into more imaginary lines, called latitudes and longitudes. Latitude lines circle Earth above and below the equator. Longitude lines circle Earth through the North and South Poles.

◀ *Latitude lines are used to measure Earth from north to south. Longitude lines are used to measure Earth from east to west.*

# What Makes Earth's Seasons?

Lean your head to one side. Your head is tilted. Earth's axis also has a tilt. This tilt causes Earth's four seasons. As Earth circles the Sun, the tilt causes one of Earth's Northern and Southern Hemispheres to be tilted toward the Sun. The other is tilted away. In the hemisphere that tilts toward the Sun, it is spring and then summer. In the hemisphere that tilts away, it is fall and then winter. As Earth moves around the Sun, the hemispheres' positions toward the Sun change. This causes the seasons to change, too.

*In this picture the Northern Hemisphere would have spring, then summer. The Southern Hemisphere would have fall, then winter.* ▶

Northern Hemisphere

EARTH

Equator

Southern Hemisphere

SUN

# Why Do Scientists Study Earth?

Scientists study Earth to learn about its past and its future. Some scientists study Earth's layers. Others want to learn more about Earth's atmosphere. Scientists also study Earth's earthquakes, volcanoes, and oceans. Scientists use different tools to study Earth. They use **satellites** that orbit Earth. Satellites take photos of Earth's atmosphere. Others study Earth's changes with samples of its soil and water. What they find will help everyone learn more about Earth. It may also help us to take better care of our planet.

◀ *Scientist Kevin Hadley sets up a tool to measure the flow of mud near Cotopaxi Volcano in South America's Andes Mountains.*

# How Can You Find Your Way on Earth?

To find your way around Earth, you can use maps and **globes**. Places on maps and globes are drawn smaller than they really are. A scale shows just how small. The **key** shows symbols that are used. Symbols are pictures that stand for other things. Maps and globes also have cardinal (N, S, E, W) and intermediate (NE, NW, SE, SW) directions. These directions are drawn on **compasses**. Maps may also have **locators**, or smaller maps on larger maps, to show towns, states, or countries in greater detail.

# Glossary

**atmosphere** (AT-muh-sfeer) The gases that surround a planet.

**axis** (AK-sis) A straight line on which an object turns or seems to turn.

**compasses** (KUHM-puhs-ez) A picture on a map or globe that shows north, south, east, and west (N, S, E, W).

**equator** (ih-KWAY-tur) An imaginary line that circles Earth's middle.

**galaxy** (GA-lik-see) A large group of stars and planets.

**globes** (GLOHBZ) Round models of Earth that spin on poles.

**gravity** (GRA-vih-tee) The force that pulls all things on Earth toward Earth's center.

**hemispheres** (HEH-muh-sfeerz) Halves of a ball, or sphere.

**key** (KEE) A list that explains the pictures or words used to represent other things.

**locators** (LOH-kay-terz) Maps inset on larger maps to show detail.

**magma** (MAG-muh) Rock that is so hot it has turned into liquid.

**meteoroids** (MEE-tee-ur-oidz) Rocks that orbit planets.

**ozone** (OH-zohn) A layer of oxygen gas that surrounds Earth.

**satellites** (SA-til-yts) Spacecrafts that orbit Earth.

**temperature** (TEM-pruh-cher) How hot or cold something is.

# Index

**A**
atmosphere, 5, 21

**C**
Columbus, Christopher, 13
compasses, 22
core, 6
crust, 6

**E**
equator, 17

**G**
globes, 22
gravity, 9–10

**H**
hemispheres, 17–18

**L**
latitude, 17
longitude, 17

**M**
mantle, 6
maps, 22

Milky Way galaxy, 14
Moon, 13

**N**
North Pole, 10, 17

**P**
Pythagoras, 13

**S**
satellites, 21
South Pole, 10, 17
Sun, 5, 9, 13–14

# Web Sites

To learn more about Earth, check out these Web sites:
http://kids.msfc.nasa.gov/Earth
http://starchild.gsfc.nasa.gov/docs/StarChild/solar_system_level1/earth.html